George Thornton Pett

The Ceylon Tea-Makers' Handbook

George Thornton Pett

The Ceylon Tea-Makers' Handbook

ISBN/EAN: 9783337230166

Printed in Europe, USA, Canada, Australia, Japan

Cover: Foto ©Lupo / pixelio.de

More available books at **www.hansebooks.com**

THE

CEYLON TEA-MAKERS'

HAND-BOOK.

COMPILED BY

Geo. *THORNTON PETT.*

PRICE Rs. 2·00 NETT.

COLOMBO :
PRINTED AT THE "TIMES OF CEYLON" STEAM PRESS.

1899.

CONTENTS.

INTRODUCTION - - - - - - - - 1

PART I.

GENERAL DUTIES OF TEA-MAKERS - - - - 2
GENERAL DUTIES OF WITHERING COOLIES - - - 6
GENERAL DUTIES OF ROLLING COOLIES - - - - 6
GENERAL DUTIES OF ROLL BREAKER COOLY - - - 7
GENERAL DUTIES OF SORTING ROOM COOLIES - - - 7
GENERAL DUTIES OF FIRING COOLIES - - - - 7
GENERAL DUTIES OF ENGINE DRIVERS - - - - 8
GENERAL DUTIES OF WATCHMEN - - - - - 9

PART II.

MANUFACTURE - - - - - - - - 10
REMARKS ON WITHERING LEAF - - - - - 10
REMARKS ON ROLLING LEAF - - - - - 15
REMARKS ON FERMENTING LEAF - - - - 17
REMARKS ON FIRING LEAF - - - - - - 19
REMARKS ON SORTING AND TASTING TEA - - - 21
REMARKS ON BULKING, FINAL FIRING, TAREING
    AND PACKING TEA - - - - - 23

PART III.

MACHINERY - - - - - - - - - 28

WATER WHEELS, PELTONS AND TURBINES - - - 28

STEAM ENGINES, MANAGEMENT, METHOD OF WORKING

    AND CARE OF - - - - - - - - - 29

OIL ENGINES - - - - - - - - 41

FANS - - - - - - - - - - 45

ROLLERS - - - - - - - - - 46

ROLL BREAKERS - - - - - - - - 48

OXIDIZERS - - - - - - - - - 48

DRIERS - - - - - - - - - - 49

SORTERS - - - - - - - - - 60

CUTTERS - - - - - - - - - - 61

PACKERS - - - - - - - - - 61

PART IV.

USEFUL NOTES FOR TEA-MAKERS - - - - - 63

THE CEYLON TEA-MAKER'S HAND-BOOK.

INTELLIGENT Tea-makers have often expressed to the writer their desire for a set of general rules for their guidance with the methods of using the various machines employed.

This little book is an attempt to comply with the wish, and is published in the hope that it may be useful to that hard-working body of men—the Tea-makers of Ceylon.

It has no pretension to teach Tea-making to experienced planters, but aims simply to be what its name denotes, a handy book for the Tea-maker. The information given has been gathered from the scattered books and pamphlets on the subject current, together with some results of the compiler's experience gained whilst planting for eighteen years in Ceylon, during eleven of which he has had to do with the manufacture of Tea.

Mr. M. KELWAY BAMBER has most kindly permitted him to make extracts and generally utilise the valuable information contained in his book on "*The Chemistry and Agriculture of Tea*", and he is also indebted for many useful hints and suggestions to his brother, Mr. F. W. A. PETT.

I

PART I.

Sec. 1.—General Rules for Guidance of Tea-makers.

(1) On arrival at factory in morning to inspect leaf and to turn on or off fans as may be necessary.

(2) To see that all withered leaf is taken off tats, weighed and taken to shoots above rollers.

(3) To see that all leaf not ready for Rolling is turned over on the tats. The reason for this is given under paragraph "Withering" in Part II.

(4) When there is an Engine, to see that it is clean, test waterguages and cocks and see that steam is up.

When there is water power to send a cooly along the water course to see that it is in order. To inspect water wheel, pelton, or turbine to see that they are clean, the bearings oiled and lubricating cups or bottles full and in working order.

(5) To inspect rollers and roll-breakers, to see that they are thoroughly clean and the bearings oiled.

(6) To see that shafting lubricating bottles are full, and that the hangers or brackets are firm—if any shake they should be tightened up—and to see that the oil catchers are in position.

(7) To see that Sifting Room is clean and that Sifters and Cutters have been cleaned and the bearings oiled. To point out Tea to be graded to sorting coolies and to start them at work.

(8) To see that Driers are thoroughly clean, including smoke flues, furnaces and ash pits, and where Desiccators are used to see that the wire mesh over air ports is free from dust ; where Down-Draft Siroccos are used to see that the wire web screen in air chamber has been thoroughly cleaned.

(9) Rolling time tables should be kept by the rolling coolies, but time, pressure, &c., should be frequently checked by the Tea-maker.

(10) To inspect fermenting leaf, and to decide when it should be finally rolled and fired.

(11) To see that firing coolies keep the temperature of the machines at the degree ordered, and that stoking is properly done.

(12) To frequently inspect the fired tea as it leaves the Driers.

(13) To inspect sorted tea, draw samples, weigh up, and ascertain percentages when day's sifting is done.

No sorted teas should be put into bins until tasted and passed. It is advisable that the Tea-maker should taste daily the teas he is making and permission should be obtained for this. A separate set of cups, &c., is often provided for the purpose.

(14) To weigh up green leaf when it arrives, direct Withering Coolies how and where it should be spread, and enter quantities in leaf book. It is useful to affix boards to the tats showing what leaf is spread on them. Chalk can be used for this as it can be rubbed out easily day by day.

(15) At 5 p.m. to see that all leaf received in factory in morning and early afternoon is turned over on the tats.

(16) To serve out shooks, lead, solder, hoop iron nails, &c., as required and to enter quantities at once in Store Journal.

(17) When boxes are lined and before Tareing to carefully look over every one to see that there are no holes in the wood, that lead lining is intact, free from holes and properly soldered. If any should be found defective, to point out what is wrong to packing cooly and inspect same again before they are tared. Before packing, scales should be tested with Standard weights and adjusted.

(18) To tare all boxes with lid, lead cover, nails and hoop iron, as directed in Bulking Regulations given further on.

(19) To number boxes and enter number and tare in Invoice Book. To carefully supervise all bulking, refiring, packing and weighing himself, and after lead cover is soldered down, to inspect lead top to see that there are no holes in it, that soldering has been properly done, and any excess of soldering fluid wiped off. Carpenter should be instructed not to nail down lid until box has been ticked off as correct by Tea-maker.

(20) To check numbers and marks on boxes with Invoice Book before the boxes are despatched from the Factory.

(21) To see that all fire buckets are kept full of water and that they are at their right stations. If there is a hose, to periodically test it by turning on water. If fire grenades, to see that they are in their proper places and the number correct.

(22) To have Factory windows cleaned as often as possible.

(23) To see that lamps downstairs are kept properly clean, the wicks trimmed, and filled with oil. To take particular care that no oil lamps be used in withering lofts or near tats. Candles in cheap bazaar lanterns only should be used upstairs ; as if capsized or knocked over the mere act of falling will generally put them out, if not a cooly can at once stamp them out without danger to himself or the factory. Whereas if kerosine oil lamps are used and should get knocked over while alight, the oil will at once burst into a flame and probably ignite the surroundings before it can be smothered.

(24) As work is finished for the day, to see that all machines are cleaned, except furnaces, which must be left to cool down, and to have factory swept out. Before leaving factory to see that all lights and fires are extinguished.

(25) To weigh up all teas in store at end of each month.

(26) To keep a careful account of all fuel—wood, coal, or oil as it is received and used daily.

Tea-makers should so endeavour to arrange their wither that all leaf is worked off during daylight, as it is difficult to make good tea at night.

In every factory each cooly should have his own definite or detailed job. It is false economy to take the nearest cooly for every little thing that turns up and to be constantly shifting the men from one work to another.

No cooly should be allowed to wear a hanging cloth or streamers from his head ; such are liable to be caught in the belting or machinery and cause accidents. A good plan which is adopted in some factories is to supply the coolies with a uniform consisting of short breeches, coat and small round cap.

When such is provided coolies should change before they begin work in the morning ; hooks should be provided for them to hang their own clothes up on.

No cooly should be allowed to leave the factory during working hours without first obtaining permission from the Tea-maker.

Sec. 2. General duties of Leaf-spreading or Withering coolies.—If leaf is ready in the morning, to remove from tats as directed by Tea-maker and turn over on tats all leaf that is not ready withered ; to spread leaf when it arrives, where and as ordered by Tea-maker. To sweep up and clean floor when leaf has been removed from tats.

Head withering cooly to put on or off fans as directed by Tea-maker, to keep fans, standards and countershaft clean and all bearings well oiled.

At 5 p.m. to turn over on the tats all leaf spread during morning and early afternoon. Method of turning over is given in paragraph headed "Withering" in Part II.

Sec. 3. General Duties of Rolling Coolies.— In morning to oil all bearings, and commence rolling when ordered by Tea-maker : rolling to be done in manner, as regards pressure and time, as directed. To take roll to and from Roll Breaker assisted by the cooly who works that machine. When rolling for the day is finished, coolies in charge of Rollers to thoroughy wash down their machines. All brass and bright steel work about the machines should be cleaned with Brooke's soap or other efficient substitute.

Sec. 4. General Duties of Roll Breaker Cooly.—In medium sized Factories this man is usually put in charge of the line or main shafting and is responsible for its brightness. In morning he should fill up all oil bottles on shafting, oil the bearings of Water Motor if there is one, and those of the Roll Breaker. He also attends to the fermenting leaf ; directions for this are given in paragraph headed "Fermenting." When day's work is done he should thoroughly wash down and clean his machine, particularly the mesh of the sieve.

Sec. 5. General Duties of Sorting Room Coolies.—In morning to oil bearings of machines, replenish oil in lubricating bottles, and clean shafting in Sorting Room. To grade tea as directed by Tea-maker, and when finished clean machines, sweep floor, dust bins and windows, and arrange the sorted tea in order for inspection. When tea is tasted, sorting passed and weighed up, to put in bins.

Sec. 6. General Duties of Firing Coolies.—In morning to thoroughly clean Driers, taking out all dust and broken tea from interiors, clean smoke flues, furnaces, and ash pits, and then oil all bearings. When Desiccators are used to carefully examine daily in the morning the packing round the smoke flues, and if found in any way loose or defective it should be replaced and forced in with a chisel, asbestos cement being also applied if necessary. With Siroccos the interior of the furnace should also be inspected to see that the packing round the radiator plate is air tight. To fire teas at temperatures ordered by Tea-maker. Temperature can be kept steady by careful stoking. Stoking directions are given in paragraph headed "Siroccos," Part III.

Fired tea should be taken to the Sorting Room

in open boxes covered with a cloth and left till morning. It should not be put into Zinc bins until it has been sifted or is quite cold ; reason for this is given under " Firing," Part II.

Sec. 7. General Duties of Engine Drivers. The following short rules have been kindly supplied to the writer by Messrs. Brown & Co., Ltd., Hatton, but the management and care of Engines and Boilers will be found to be more fully treated on in Part iii—"Machinery."

(1) Warm boiler gradually. Do not get up steam from cold water in less than four hours.

(2) Moderately thick fires are most economical. Fire evenly and regularly, a little at a time. Do not clean fire oftener than necessary, and keep fire door open as short a time as possible.

(3) Cleaning must be done thoroughly inside and out. This frequency of cleaning will depend on the nature of fuel and water ; but the boiler ought to be opened at least every two months.

(4) Never fill a hot boiler with cold water.

(5) The dirty water should be blown off every morning ; allow the cock to stand open for two or three minutes when the steam pressure is about 5 lbs.

(6) If the boiler is not required for some time, fill full of water containing a quantity of common washing soda ; or fill nearly full and pour on this a quantity of crude petroleum, and then run out water.

(7) Guage cocks and water guages must be kept clean.

Water from guage glass should be blown at least twice a day. If the water does not return quickly to the glass the connections require cleaning, which can be done with a wire. It does not follow that there is plenty of water in the boiler, because it shows in the glass, hence the importance of blowing through the guage cocks frequently.

(8) Lift each safety valve by hand in the morning to see that it is free.

(9) Do not empty the boiler under steam pressure, but cool it down with the water in, then open blow out cock.

(10) Check valves and self-acting feed valves should be frequently cleaned. Get the feed valves so as to give a constant supply and keep the water up to say half glass.

(11) In case of low water immediately cover the fire with ashes and earth, wet if possible, and open furnace door. Draw fire as soon as it can be done without increasing the heat. Never turn on, feed, start, or stop the engine or lift safety valve, but let boiler cool.

(12) The principal points to be observed in the care of engines are to keep all wearing parts well oiled, and in thorough repair, and thoroughly clean; everything about an Engine and Boiler Room should be kept clean and tidy; dirt increases the wear and tear and often hides faults which would otherwise be noticed.

(13) Should engine not be required for a short time, the fly wheel should be turned through one or two revolutions every day.

Sec. 8. General Duties of Watchmen.— They should, when work is finished in the Factory, go round and see that all locks and windows in ground floor are fastened, also that the doors of all furnaces outside the factory have been closed or the fires drawn. In some Factories where the Fans are kept going at night it is part of their duty to see that Water wheel, Turbine, or Pelton is working properly, that the bearings do not get heated, and, if necessary, to replenish the oil in the lubricating bottles on same.

PART II.

⤚⇛‧ M A N U F A C T U R E . ‧⇚⤙

In all the processes of manufacturing tea, Tea-makers must of course follow out the instructions they receive from their Masters ; but the following remarks on the subject may be found useful :—

Sec. 1. Withering.—When leaf arrives at the factory it should be carefully examined by the Tea-maker, and if found at all bruised or heated from being jammed, or kept for too long a time in bags or baskets, the fact should be reported to the Superintendent, and, if possible, the damaged leaf should be withered and made separately. Such leaf turns red, and where fractured black, or as Mr. Bamber expresses it, " a process of oxidation " or decomposition sets in almost immediately " and causes a loss in the appearance and quality " of the tea." In other words, there will be a large proportion of red leaf in the tea and great loss of flavor.

The object of withering is to get rid of the superfluous water and render the leaf fit or pliable for rolling ; properly withered leaf will retain and keep a good twist without being broken when rolled.

Leaf should always be spread evenly ; it will not do to have some tats lightly and some thickly spread. Coolies are apt to leave small heaps here and there on the tats, unless well watched by head loft cooly. The degree of thickness that leaf should be spread must be regulated by the weather, withering space and capacity of rollers available. Leaf brought in during the forenoon should not be spread very thin in dry weather, or it will be ready for rolling before daylight the following morning. To aid an even wither, leaf spread during the forenoon or early afternoon should all be turned over on the tats about 5 o'clock p.m., and the next morning the leaf spread on the previous evening should be treated in the same way. The reason for this is that the breathing spaces of the leaf are nearly all on the under side, and the moisture will of course be more quickly evaporated when this is uppermost or in direct contact with the air. The chances are that by turning over the leaf after a few hours a large proportion of it will be evenly affected. To turn the leaf coolies should strike the underside of the tats gently with light rods ; this is better than turning the leaf from above, and coolies soon get very expert at it. It is generally found that leaf withers better on wooden tats than on hessian ones, and Tea-makers should bear this in mind when the factory is provided with both, so as to have a regular supply of leaf for the rollers. When the leaf comes in very wet, it is desirable to get the moisture *off* it before spreading for withering, and if the factory is provided with Fans this is easily accomplished in an hour or two, by spreading it not more than three inches thick on tats, close to them, and having it turned by hand say every quarter of an hour ; when the water has all disappeared the leaf should be taken down and spread in the ordinary way. Very wet leaf always has a hard look and is brittle, the result, it is supposed, of the cells being overcharged with water, and some Tea-makers are apt to at once think it is coarse. If, however, it is treated by the Fans in the way I have mentioned, its real quality is readily ascertained. Very wet leaf should be rather more withered than usual to concentrate the weak sap.

Should there be only a small amount of leaf brought in for the day, as sometimes happens in most factories, it can be worked off economically by getting the morning and evening leaf withered at the same time ; to do this the evening leaf should be spread only a third as thick as the morning leaf. In ordinary weather, should a quick wither be required, it can be obtained by spreading the leaf evenly and so that no one leaf overlaps or touches another on the tats. Coolies can soon be taught to spread the leaf in this way, and it is the way "fancy teas" are withered.

A very speedy wither for a small quantity of leaf can nearly always be ensured by spreading the leaf in the same way, i. e., no one leaf overlapping another, on the wooden floor of the loft below the tats.

Tea-makers should be most careful to see that no leaf is left on the floor in spaces between the rows of tats or gangways, as all leaf trodden on and bruised, more especially when it is partially withered, turns red and spoils the tea.

Well withered leaf is soft and flascid to the touch, and when rolled up into a ball and lightly pressed retains its shape ; the stalks too, on being doubled over, should bend easily without breaking. Really well even withered leaf feels like a silk handkerchief and gives out a fresh almost pleasant smell, quite different to the mouldy vegetable smell of badly withered leaf. Uneven withered

leaf contains black, brown, red, and green leaves
with stalks that break on being bent. Leaf should
not be withered in the sun, if it can be possibly
avoided. Leaf withered in the sun gives red tea.
Leaf should never be left about the lofts in heaps
as it turns sour.

An under wither is to about 64 per cent, that
is each 100 lbs. of green leaf is withered down to
64 lbs.

A light wither is to 58 to 60 lbs., called 58 or
60 per cent.

An ordinary wither is to 53 to 55 lbs., called
53 or 55 per cent.

A hard wither is to 48 to 50 lbs., called 48 or
50 per cent.

The above calculations are for leaf gathered and
brought into the Factory, in ordinary dry weather ;
of course if the leaf is wet the calculations will
vary with the degree of moisture on it.

Under withered leaf breaks in the Roller ; the
juice from a handful comes with a slight squeeze,
is watery and of a light greenish milky colour.
The tea is of a reddish grey colour, and the liquor is
very light in colour, cloudy, weak, soft and tasteless.

A light wither breaks in the roller also, though
not to the same extent ; the juice expressed is
a little darker, and so is the liquor, which is weak
but pungent. Roll colours quickly but not evenly.

An ordinary wither gives the best results,
at medium elevations, say from 2,500 to 4,200
feet above sea level. Leaf takes a good twist,
juice is dark mahogany colour, and requires
a good hard squeeze to extract it; roll colours
quickly and evenly, if properly attended to, and
infused leaves are bright and coloury, if the
succeeding processes have been carefully carried out.

A hard wither is apt to break in the Roller, seldom colours during fermentation, and the infused leaves are dark greenish ; liquor is darkish and inclined to be mawkish. No juice can be pressed out by the hand. A hard wither, however, is often adopted, with good results, at high elevations, when the tea is to be fired without fermentation. Hard withered leaf is said to give most tips, but I have not found it so.

Badly over withered leaf stinks in wet weather and should be thrown away.

Tippings, though containing a great deal of moisture, wither quicker than ordinary leaf.

It some times happens that during a drought a hot wind prevails during the night, which drys up the leaf in spite of closed windows, and no matter how the leaf has been spread on the tats. This dry leaf, it should be noted, is not necessarily withered leaf, as can be seen when the tea made from it is infused for tasting. When leaf gets dry as indicated, the best thing to do is to fill up the Rollers half full with it, pour in half a bucket of water, then start the rollers at a slow speed without pressure from the top, add the rest of the leaf, and pour in a little more water as required, as the rollers get a grip on it. It will then take a twist instead of being smashed up into flakes or ground to powder.

Leaf can only be properly withered with dry air. Windows of withering lofts should always be open unless there is a too dry wind (in which case those on the windward side should be closed and those on the leeward opened), wet weather, or the Fans are at work. In any case if windows have been closed during the night and there are no Fans, some of them at any rate should be opened in the morning to ensure a circulation of air and to carry off the moisture accumulated in it from the leaf during the night.

Sec. 2. Rolling.—The object of rolling is to twist and then break up the *cells* of the leaf, without actually breaking or fracturing the leaf itself. The cells contain juice or sap, and by breaking them up the juice is liberated or set free, to be diffused or spread over the outer surface of the leaves.

Properly withered leaf takes a good twist without being smashed up or broken.

Lightly rolled leaf gives a tippy tea with but little strength.

Hard rolled leaf gives a strong wiry tea. The tips are discoloured, and if the leaf has been either over or under withered the tea will be choppy or broken up.

Leaf from bushes a long time from pruning requires a harder rolling than ordinary leaf. Leaf from bushes in seed, or flower, will however have but little strength, no matter how hard it is rolled ; it will also be lacking in flavor, probably from most of the sap of the bushes being taken up in the effort of reproduction. This leaf too is generally very banghy, probably from the same cause. Coarse leaf wants as hard a rolling as it can get.

Tippings or leaf from recently pruned bushes take a light roll, as the leaf is soft. It is impossible to make a strong tea from these, as the sap is immature and watery; it is therefore better to make them for appearance, by medium wither, light rolling, and brisk firing.

The ordinary method of rolling adapted in many Factories is to roll the leaf three times in periods of half an hour each ; the roll being sifted through the Roll-Breaker at end of each half hour, The small leaf that falls through the sieve of the Roll Breaker is taken away, and the large returned to the roller. The time in Roller should be counted from time all the leaf for a roll is in the machine, Leaf is generally put into the Rollers down a shoot from the upper floor ; when the box is about a third full, the Roller can be started, and the rest of the leaf for the roll let down by degrees as the machine can take it in. Rollers should never be jammed too full. Amounts of leaf and speeds to be driven are given in Part III. paragraph Rollers. When leaf is being rolled the first time no pressure should be applied, but the lid of the top of the box must be lowered to just feel the leaf. If pressure is applied too soon, the leaves do not take a twist but are smashed up and flattened, and the made tea is flaky and choppy and all tips discoloured. When rolling for second time pressure should be gradually applied by lowering the lid during the first seven minutes ; it should then be raised high up for three minutes, to enable the leaf to get a good turn over, the air will thus have free access to it and keep it cool. I find it best to subdivide each half hour into three periods of ten minutes each, keeping pressure on for seven minutes and raising lid for three minutes, to turn and cool leaf, &c.

Rolling coolies can easily be taught to do this. When lid is raised, Roller cooly can assist the turn over by breaking up and handling all lumps that come up at feed-hopper. When rolling for third time, if *hard* rolling is required, the top can be put down as far as possible, care being taken though to raise it for three minutes out of each ten to allow it to cool.

In some factories leaf is only rolled twice, *i. e.* in two periods of forty-five minutes each. The operation is usually the same as that mentioned above, except that pressure is gradually applied after the leaf has been in the Roller for twenty or twenty-five minutes the first time. Well rolled leaf is closely twisted, gummy to the touch, and has a fresh smell. The large leaves after infusion, if held up to the light do not have a blotchy, patchy appearance, but are of an even coppery-brown colour with well marked lines all over. Over rolled leaf, *i. e.* leaf hard rolled for too long a time, say over two hours, makes a soft tea, and the flavor is nearly all lost.

Leaf is often rolled for a fourth time for a few minutes before it is put in the Drier ; in this case no pressure whatever should be applied, the lid of the box being lowered to only just touch the leaf, the object being merely to re-twist any leaves that may have opened out during fermentation, and to excite a little moisture which aids an even colour. This fourth or final rolling is a very good thing to do when possible. I do not consider it necessary to re-roll the small leaf that has gone through the Roll Breaker sieve, as it discolours the tips. It is, however, sometimes done for about ten minutes, and sometimes it is lightly re-rolled by hand to give it a little more twist.

Sec. 3. Fermenting,—or properly called Oxidation. This is a very important process in tea manufacture, and requires the constant attention of Tea-makers. It should be carried out in the coolest part of the Factory, or at any rate at a distance from the Driers, as it is impossible to obtain a good colour if the room is too hot.

No fixed time can be laid down for this process. I have known leaf attain a bright even coppery colour ten minutes after it has left the roller, and on other occasions I have known it to take up

to three hours. The time taken and the colour
are affected by the quality of the leaf, temperature
of the air, and the extent to which the leaf has
been withered.

Slightly underwithered leaf colours quickly and
evenly ; well withered leaf colours evenly but not so
quickly ; hard withered leaf seldom looks coppery
coloured, but turns dark greenish and if left long
enough, black.

Mr. Bamber in his book (already mentioned)
has the following on the subject which bears out
Ceylon experience :—" When the leaf is first
" brought from the Rollers it is of a bright green
" colour (if properly withered); but after lying
" for about half an hour under favourable con-
" ditions of temperature and moisture, it begins
" to gradually assume a reddish tint, especially
" near the spots where the leaves and stems are
" fractured. This change continues until the
" younger leaves and stems are a bright coppery
" colour, while the older and less perfectly
" rolled leaves are partly reddish and partly green."

At medium elevations in Ceylon leaf begins to
colour in the Rollers if not over-withered. If the
leaf has been rather over-withered and is at all dry
when it leaves the Rollers it should be moistened,
not saturated, with clean water whilst it is being
spread to ferment ; this is best done by spraying it
lightly with a garden syringe having a fine rose.
Leaf for fermentation is generally spread on a cement
floor (though sometimes on wooden trays, tats
or tables, it however does better on cement) about
four to five inches thick, and kept covered with a
damp cloth. In very hot dry weather the cloth
should be kept thoroughly wet, and if the leaf has
been at all over-withered, it can be moistened
with a little water from the syringe each time it

is turned. A thermometer should be inserted in the leaf under the damp cloth, and whenever the temperature rises to 85° the leaf should be turned over. Under certain conditions the temperature does not rise as high as this, and on these occasions the leaf should be turned over every half-hour. Well fermented leaf should look bright and coppery at first glance on taking up a handful. The larger leaves will still look greenish if the handful is carefully looked into, but it is not advisable to wait until all these are copper coloured, or the finer leaves will be overdone. Well fermented leaf has a fresh sweet nutty smell. In some Factories the leaf is fired almost as soon as it has left the Rollers for the third time and entirely unfermented. If it is at all dry, say from being overwithered, it is advisable to moisten it with a fine spray of clean water before putting it into the Driers.

As mentioned under paragraph "Rolling", the leaves open out a little while undergoing fermentation, and are often put back into a Roller and rolled for a few minutes (five to ten) without pressure just to retwist them immediately before they are put on the Driers.

Sec. 4. Firing.—The main object to be aimed at in firing is, of course, to remove all the moisture without driving off and losing any of the constituents which add to the flavour and consequent value of the tea (*Bamber*). Properly fired tea has an excellent aroma when it leaves the machine ; it is crisp to the touch and slightly springy,

It is known that fermentation goes on in the leaf after it has been put into the Drier ; to check this it is generally put in when the latter is at a high temperature. Many of the machines now in use are specially designed to check this, and details of the way to work them will be found in Part III, Paragraph Driers.

It should be noted that the colour of the infused leaves when being tasted will be about the same as that of the fermented leaf when put in Drier, provided that the fermentation is checked at once. The use of a high temperature throughout the firing is now generally avoided. As after further fermentation is arrested it causes the evaporation of one of the chief flavouring constituents, an essential oil, which combines with the moisture carried off by the Fan and is thus lost. In some Factories where there are no machines that work as those indicated, a system of "half firing" is employed ; the fermented leaf being passed through the Drier quickly at a temperature of from 230° to 240° and taken out when just about half fired. If only one machine is available the fermented leaf is put through in this way, and as the tea arrives from the Drier half fired it is put on one side until the whole lot for the day has been through, then the temperature of the Drier is reduced to 180° and the half fired tea is put through again to finish it off. Half fired tea can be kept in that state for a few hours without injury. When two Driers are available the fermented leaf can be put through one at a temperature of 230° to 240° and then passed on and put through the other at a temperature of from 175° to 185°. These processes are called "firing" and "re-firing," the last firing before packing being termed "final firing."

Fired teas should not be put into Zinc bins while they are still hot from the Driers, as though the leaf may feel perfectly dry to the touch there is always some amount of moisture present, some of which is apparently given off as the tea cools down. If then hot tea is put into closed Zinc bins it is obvious that this moisture will be retained. I have even found the Zinc sides slightly wet or clammy when tea has been put into bins thus. I find that it is better to empty the trays of fired tea into open wooden boxes, take to Sorting Room, cover with a light cloth, and leave till morning, when the tea will be found ready for sifting.

DRY fuel should always be used for firing, as if used wet or sodden a great waste is caused, as much of the heat generated by the combustion or burning of a portion is taken up in drying the remainder and rendering it fit to burn. It is most important that the temperature of the air in the Driers should be kept steady, and this can only be done by using dry fuel.

A good deal of information on Firing will be found in article Driers, Part III.—Machinery.

When tea is to be fired it should be spread evenly and thinly on the trays or webs. If spread too thick, which coolies are fond of doing, the made tea will be stewed, taste mawkish, irregularly fired, and take longer to go through the machine.

The interior of the Driers must invariably be thoroughly cleaned every morning before commencing work, as should any broken tea, &c., be left in, it will probably get burnt when the machines are again used and then mixed with the following day's make.

Sec. 5. Sorting.—Tea can be sorted into so many different grades and in such various ways that it is useless here to formulate any definite rules on the subject. A few remarks are, however, given, and more will be found in Part III, paragraph *Sifters*.

It may be stated generally that leaf that won't pass through a No. 4 mesh sieve is Red, Congou, and Souchong.

Above No. 8 sieve there remains Pekoe Souchong, above a No. 10 sieve Pekoe, above a No. 12 Orange Pekoe, above a No. 14 Broken Pekoe, and what passes through a No. 14 Broken Orange Pekoe, from which the dust is generally taken by putting it over a No. 28 or 30 sieve. Flat leaf taken from the different grades by a hand winnower is called "fannings", and by some "broken mixed". Tea, however, is seldom sifted into so many grades as those enumerated above, and Superintendents are usually guided in their selection by the varying requirements of the market and also by the style of leaf they are plucking.

Teas should be tasted after sorting and before they are put in the bins, and it is advisable that the bulk and small tea of the previous day's make should be tasted on the following morning, so that anything wrong in the manufacture can be detected at once.

Tasting. The following hints on tasting have been abridged from the Tea Planters' Vade Mecum and Rutherford's Note Book.

In tasting tea the Tea-maker should look to the four following characteristics :—

    (1)   Its nose or smell.

    (2)   Its liquor.

    (3)   Its Infusion.

    (4)   Its leaf.

  (1)   Its nose, whether strong, **rich scent, minty, nutty, musty, burnt, high fired,** or **brisk.** Judged by smell.

(2) Its liquor, whether strong, rasping, pungent, brisk, flavoury, full thick, malty, dark, or wanting in strength, dull, insipid, thin, burnt, overfired, soft. Judged by taste.

(3) Its infusion, bright or dark coloured or mixed with green or any dark or burnt leaves, amount of stalk, and whether over or under fermented. Judged by sight.

(4) Leaf. Its make and appearance, whether black, wiry, even, regular, well, little, or open twisted, hard or light rolled, flaky, bold, tippy, grey, dusty or irregular, wanting in tips, &c. Judged by sight.

Sec. 6. Bulking, Final Firing, Tareing and Packing.—Bulking is done in various ways, the object being to thoroughly mix up the tea to be packed, so that a handful or sample taken any where from the mass represents exactly the quality of the entire lot. The operation should be carefully performed. The tea to be bulked is taken from the bins and piled in a heap on a bulking cloth spread on floor of Packing Room ; the packing coolies then surround it and pull down the tea from the centre and throw it outwards on the cloth, leaving an empty space in the middle ; it is then thrown back by shovels or hand into a pile again ; this process is generally done three times. Another way of bulking is to take the tea to an upper floor and pour it down a hopper or shoot on to a cloth on the floor of the packing room when it should be well turned over and mixed by shovels or hand. The tea is then final fired.

Final Firing.—The object of final firing is to extract any moisture that may be remaining in the tea, and it should be noted that tea feeling quite dry and even brittle to the hand may contain as much as from 10 to 15 per cent of moisture. Tea is usually final fired at a temperature not exceeding 180 degrees; if a higher temperature is employed there is a risk of losing flavour by the evaporation of an essential oil, as explained before. The tea should be passed through the Drier *slowly* to extract as much moisture as possible. In some Factories final firing is done at a temperature of about 220 degrees. This renders the tea brisker, but I question if this briskness is fully retained, say until the tea is sold in London. Tea should be allowed to cool down a little after final firing, and should be packed when it is *warm, not hot.* If the weather is at all wet when packing is going on, all the windows in Packing Room should be kept carefully closed, to exclude damp air, which would be readily absorbed by the tea.

As Mr. Bamber says: " The object of packing tea " warm is to prevent, as far as possible, the absorp- " tion of moisture from the atmosphere, which " might tend to promote further undesirable " change in the tea during a long voyage. Tea, on " the other hand, should not be packed *hot,* " although it would ensure the absence of moisture " at the moment of packing, for if it is, the air in " the box will be heated and expanded to such a " degree when the lead is soldered down, that on " cooling a partial vacuum would be formed, and " this would cause an inward passage of air through " the slightest defect in the lead."

Tea is now generally packed with a machine, and the process is described in Part III. paragraph **Packers.**

Where there is no machine it is usually slowly poured into the boxes, which are rocked over a reeper or rounded stick until they are filled with the desired quantity. Tramping in with the feet is most objectionable, and is apt to break up the

THE LONDON CUSTOMS SYSTEM OF WEIGHING CEYLON TEAS UNDER THE NEW REGULATIONS EXPECTED TO COME INTO FORCE ON OCTOBER 1st, 1899.

INSTRUCTIONS SHOWING HOW TO PREVENT A LOSS ON GARDEN WEIGHTS.

At the settlement of the recent dispute between Growers and Buyers. the Joint Committee of the Growers' and Buyers' Associations arrived at the following decision with regard to the *Weighing of Tares :*—

" If the empty package weighs an even pound it is to be entered as such ; if it weighs an even half-pound or over it is to be entered as the next pound above, and if it weighs less than half-a-pound it is to be entered as the pound below; boxes are to be weighed and tared as heretofore."

In order to fully benefit by the above change from the system of weighing now in vogue by the London Customs Authorities, the following hints will prove of value to Proprietors and Superintendents of Estates :—

1. The Customs will in future give the turn of the scale on the *Gross against* the Importer. but on the *Tare* either *for or against* the Importer when it weighs under or over the half-pound.

2. The *Tare* (that is the weight of the empty package complete with lid. lead, hoop-iron and nails) *should weigh four ounces over the pound*, whether the package be chest or half-chest.

3. The *Gross Weight* of a package *must in all cases weigh four ounces over the pound*, whether the package be chest or half-chest.

4. Subjoined is an example of the *correct* method of weighing a package containing 100 lbs of tea.

GARDEN WEIGHTS, CEYLON.

Tare.	Tea Net.	Gross Weight.
lbs . ozs.	100 lbs.	127 lbs . ozs

CUSTOMS WEIGHTS, LONDON.

Gross Weight.	Tare.	Net.
127 lbs.	27 lbs.	100 lbs.

From the above it will be seen that a margin of 4 ozs. remains for any slight variations in weight of package during transit, and that *no loss* need result from the Customs weighing under this system, whereas now a minimum loss of 5 ozs. per package is unavoidable.

5. Should the present system of weighing in Ceylon be continued after the new Regulation has been put in force by the London Customs, the following would be the result:—

GARDEN WEIGHTS, CEYLON.

	Tare.	Tea Net.
No. 1	27 lbs. 14 ozs.	100 lbs. 5 ozs.
No. 2	27 lbs. 9 ozs.	100 lbs. 5 ozs.

CUSTOMS WEIGHT, LONDON.

	Gross Weight.	Gross Weight.	Tare.	Net.
No. 1	128 lbs. 3 ozs.	128 lbs.	28 lbs.	100 lbs.
No. 2	127 lbs. 14 ozs.	127 lbs.	28 lbs.	99 lbs.

No. 1 shows a package correctly packed under the old system, and shows a loss of 5 ozs ; but No. 2 incorrectly packed would result in a loss to the Grower of 1 lb. 5 ozs.

6. It is most important that the weights of the weighing machine, used on the estate, should be constantly checked, and for this purpose a set of test weights should be kept. A beam scale is to be preferred to a platform one, as the former is the more accurate.

7. When a Superintendent, to equalize the tares of his package, adds pieces of lead or wood for that purpose, the material so added should be fixed inside the package, so as to prevent it falling out when the package is opened in London.

8. A Superintendent may "tare" and pack his teas with the greatest care, but, if he afterwards permits his carpenter to plane away from the top of the package before nailing down, all his careful work may be wasted.

The Metropolitan Bonded Warehouses, Limited, Crutched Fiarrs, London, August, 1899.

tea and render it quite different in appearance to what it was when it came from the sifter ; it can, however, be lightly pressed in by hand without much damage, but even this is undesirable.

Tareing. " The London Customs system of weighing Ceylon teas, and Instructions showing how to reduce the loss to a minimum," are given below.

(1)   The tare, that is the weight of the empty package complete with lid, lead, hoop-iron and nails, should in all cases weigh two to four ounces under the pound whether the package be chest, half chest, or box. *Note*—Two ounces is sufficient.·

(2)   The gross weight of a package must in all cases weigh three ounces over the pound, whether the package be chest, half chest, or box.

(3)   When a shipment of tea is not to be " Re-bulked " in London, the Customs authorities " average tare " the break, that is to say, a small percentage of the packages are opened and their tares ascertained, and from these an "average tare " for the whole break is struck. In this case it is imperative that the tare of each package weighs alike.

(4)   When a shipment of tea has to be " Re-bulked " in London the tare of each package in the break may vary *provided* the tare of each package is two ounces under the pound.

Subjoined is an example of the correct method of weighing two packages said to contain 100 lbs. tea each, which have to be re-bulked in London.

ESTATE WEIGHTS, CEYLON.

Tare.	Tea nett.	Gross weight.
No. 1 27 lbs. 14 oz. -	100 lbs. 5 oz. -	128 lbs. 3 oz.
,, 2 28 ,, 12 ,, -	100 ,, 7 ,, -	129 ,, 3 ,,

CUSTOMS WEIGHT, LONDON.

Tare.	Gross weights.	Tare.	Tea Nett.
No. 1 27 lbs. 14 oz. -	128 lbs. -	28 lbs. -	100 lbs.
„ 2 28 „ 12 „ -	128 „ -	29 „ -	100 „

The two samples above will demonstrate
the point in as much as in No. 1 ; the loss is
5 ozs. only, which is the least possible, while
No. 2 shows a loss of 7 ozs, owing to the
slightly lighter tare.

(5) The following is a very usual but incorrect
way of weighing teas, possibly through faulty
scales or weights.

ESTATE WEIGHTS, CEYLON.

Tare.	Tea nett.	Gross weight.
No. 1. 27 lbs. 3 oz.	- 99 lbs. 12 oz. -	126 lbs. 15 oz.
„ 2. 28 „ 1 „	- 100 „ 13 „ -	128 „ 14 „

CUSTOMS WEIGHTS, LONDON.

Tare.	Gross weight.	Tare.	Tea nett.
No. 1. 27 lbs. 3 oz. -	126 lbs. -	28 lbs. -	98 lbs.
„ 2. 28 „ 1 „ -	128 „ -	29 „ -	99 „

The Customs do not recognize ounces.
With regard to example I, this package, the
gross weight of which the Superintendent
makes 126 lbs. 15 oz., would only be called
126 lbs. in London ; the tare, according to
the Superintendent, is 27 lbs. 3 oz. over ; here
(*i.e.* London) the 3 oz. would be called 1 lb.
and the tare is called 28 lbs. The 28 lbs.
tare is deducted from the gross weight
of 126 lbs., with the result that the amount
of tea in this package is said to be 98 lbs , the
owner of the estate losing the 1 lb. 12 oz. tea
which may quite possibly be in the package.
A still larger loss is to be seen in example 2, in
which the Superintendent has packed 100 lbs.
13 oz. of tea, but only gets paid for 99 lbs., the
difference going into the pocket of the retailer.

(6) A most important point is to have the weights of the weighing machine used on the estate constantly checked, and for this purpose a set of test weights should be kept. A beam scale is to be preferred to a platform one, as the former is the more accurate.

(7) When a Superintendent, to equalize the tares of his packages, adds pieces of lead or wood for that purpose, the material so added should be fixed inside the package, *i. e.* between the lead and the wood at the sides so as to prevent it falling out when the package is opened in London.

(8) A Superintendent may " tare" and pack his teas with the greatest care ; but if he afterwards permits his carpenter to plane away from the top of the package before nailing down, all his careful work will be wasted.

A " sampling break " for London consists of the following :—

18 full chests or
24 half chests or
30 boxes   of each grade. Any grade containing a less number of chests, half chests or boxes is put up for sale after the " sampling " lots have all been sold, or on the following day.

For the Colombo market it is necessary that each grade should consist of at least 700 lbs. nett, if not it is included in the " small lots " and sold after the large lots have been put up.

NOTE.—In order to avoid extra charges in the London Warehouses Tea packages should weigh gross 17 to 34 lbs., 35 to 44 lbs., 45 to 59 lbs., 60 to 79 lbs., 80 to 89 lbs., 90 to 129 lbs., 130 to 159 lbs., 160 to 199 lbs. There is a separate rate for each of the series. Gross weight of packages should not vary more than two lbs. for each particular grade in the break.

sufficient water as above explained, and in case it has been filled some time before steam is required to be raised, care must be taken to ascertain that the water is the proper height, by opening both taps in the bottom water guage fitting to see that water flows freely from same, and after closing the bottom tap only, that it appears in the glass guage the right height.

Particular care must be taken to see that the taps between the guage glass and boiler are kept open at all times when the fire is alight.

Marshall's Locomotive type Boilers of 16 horse power and over have two sets of water guages, but no test cocks.

Cleaning the Tubes.—The tubes should be swept every morning or once in twenty-four hours, when at work with, a brush by passing it through each tube from the smoke box end of boiler into the fire box so as to thoroughly remove every particle of matter adhering to them. If soft coals, wood or vegetable matter is used as fuel, the tubes will require sweeping at midday or possibly at shorter intervals ; this, however, is only the work of a short time, and if the tubes are kept clean the boiler will generate steam better and with a less consumption of fuel. The smoke box should be cleaned regularly at the same time as the tubes are swept, and great care must be taken to prevent any water that may come from exhaust pipe lodging in the smoke box.

Safety valves.—The safety valve levers ought to be moved every day so as to ascertain that they are quite free in the joint. Should the steam get so high as to blow off violently from the safety valves the levers of the latter must on no account be held down by extra weights, but by pushing the fire forward against the tube plate and closing the damper on ash pan, also opening the furnace door a little, the steam pressure will gradually get reduced, and the blowing off ceases. On no account open the smoke box door when the steam gets too high.

Working Height of Water in Boiler.— During the time the boiler is in use every care must be exercised in maintaining a proper and uniform quantity of water in the boiler. This should always show in the glass tube indicator level with the water line. To prove that the guage glass indicates correctly, the two test cocks on the opposite side of the boiler front should frequently be opened, and if water and steam flow out of them this is proof that sufficient water is in the boiler and that the glass guage is in working order. Sometimes when using dirty water the passages in the brass water guage fittings may become choked up, and in such case the proper height of water may be shown in the glass, whilst there is an insufficient supply in the boiler, hence the necessity for frequently trying the test cocks above referred to.

If at any time water cannot be seen in the guage glasses, and on opening the test cocks no water is discharged, nothing but dry steam, the engine must be stopped immediately and the whole of the fire taken out, and when the *boiler has cooled down*, it must be refilled with water to its proper height before the fire is again lighted.

All the handles of glass water guage cocks, also test cocks, should be moved every day so as to prevent the possibility of their getting fast, for if they do, it will be impossible to shut off the water and steam when a glass breaks. Should a glass in a guage break, the cocks must be shut off at once at top and bottom, and the water level in boiler ascertained by means of the test cocks on opposite side, till a fresh glass tube can be inserted in the cocks.

To insert a new glass.—Unscrew the nut on top of upper cock and put in the glass with an India rubber washer on each end and screw up the two nuts, taking care that the glass is not too tightly bound up ; it ought to be free to move by hand when cold.

Lighting the Fire.—All the ashes and clinkers must be removed from the grate bars, and also from the air spaces between the bars, and likewise from the ash-pan underneath the fire grate ; it is also necessary to see that the proper quantity of bars are in the fire box and in their right position. Place a small quantity of straw or wood shavings on the grate bars, and on the top of that a thin layer of dry wood or coals if they are used, then light and add the necessary fuel from time to time until steam is raised, taking care to always keep the grate bars covered.

Firing.—Careful attention to the following rules in firing will effect a great saving in fuel and maintain a more uniform pressure of steam. The fuel, whether wood or coal, should be fed into the fire box so as to cover the entire area of the grate. The fuel should be added in small pieces, and in case the steam pressure in the boiler becomes too high, lower the damper door in front of the ash pan. The fire door should always be kept closed, except when attending to the fire, as when open it allows a current of cold air to pass into the fire box and tubes ; this is liable to cause the tubes to leak and to produce an injurious effect on the fire plates.

Cleaning out the Boiler.—It is a matter of the greatest importance to keep the boiler clean and free from any accumulation of deposit in its interior, and this must be regularly attended to. The intervals at which this cleaning out becomes necessary will depend on the water used in the boiler ; at all events it should be done after every 100 hours of working, and if the water is dirty it must be carried out oftener, say once a week.

It will be found advantageous, when using water that is not very clean, to force in an extra quantity with the feed pumps, and to raise it to a higher level than the usual working height, then open the blow off tap at the bottom of the boiler and reduce it to the proper level. The water thus blown out will carry with it undeposited sediment which would otherwise adhere to the tubes and boiler plates ; this may be done two or three times a day.

A further prevention of the adhesion of deposit to the inside of the boiler will be found in the application of common Soda. The quantity which may be used with advantage is 1 lb. for an 8 H. P. boiler, 2 lbs. for a 16 H. P, and so on in proportion per week ; the soda should first be dissolved in hot water, and the latter emptied by proportions into the feed water-tank, say one-third at a time, the soda is thereby mixed with the water and duly forced into the boiler by the feed pump ; it largely prevents the incrustation of any ingredients on the inside surface, and by periodically blowing out, as already explained, the injurious sediment is removed altogether. It is a remedy easily procured and readily applied, is very effectual, and produces no deteriorating effect on the tubes or plates. Boilers are so constructed as to be conveniently cleaned out. There are a series of small openings or " mud holes " round the bottom of outer shell of fire box, and a larger opening or " manhole " in the upper part of the boiler near to the cylinders, and by removing the covers from these openings the tops and sides of the internal fire box and the tube plate may be cleaned, and also deposit removed from between and around the tubes. The covers referred to must be removed in the following manner :—Screw off the nut, remove the cross bar, and drive the cover inwards by placing a small piece of wood on outer end of the stud, and striking same lightly with a hammer, then twist the cover across the opening and it will come out. A suitable rake should be used for removing sediment from the interior of the boiler; this should be drawn backward and forward along the sides of the fire box to remove any deposit from the plates and stays, and to rake out every portion of dirt around the bottom and sides of the boiler. A piece of hoopiron must be passed through the manhole and between the tubes especially close to the tube plate to clear away any incrustation formed **there.**

On the under side of the barrel of certain boilers there is a pocket or " collector " for the reception of any sediment which may accumulate in that part of the boiler; this is constructed with an opening or " mud hole " having a cover over it, and this latter is removed in the same way as those round the bottom of fire box ; it should frequently be taken out, and any accumulation of deposit in the barrel and around the bottom tubes cleaned out.

By unscrewing the brass plug which will be found in some boilers at the bottom of smoke box tube plate, a long rake can be passed along the boiler from the smoke box end and the dirt pushed forward, and out at the opening in the collector above-mentioned. After carefully carrying out the foregoing instructions pour plenty of clean water into the boiler at the manhole opening and swill it out thoroughly. Too great pains cannot be taken to keep the whole of the boiler as clean as possible, and it should *never* be neglected. The readiest method of replacing the covers over the manhole and mud-hole openings after cleaning the old packing off, is to place an asbestos ring round the fitting part, put the covers evenly in their respective positions, fix on the cross bars and screw the nuts up tight.

Emptying Boiler.—The whole of the fire must be taken out of the grate, and the steam pressure allowed to go down and the boiler to cool before any water is run out of it. To empty the boiler whilst under a pressure of steam would cause an irregular contraction of the plates and incur great liability to injury.

Cleaning.—The engine must be cleaned thoroughly every day it is in use; a portion may be done while it is in motion, but as the working parts can only be cleaned while it is at rest, they must be carefully attended to after it has stopped.

Oiling.—The oil used for the engine should be of the best quality. Mineral oil is far preferable to vegetable as the latter has a great tendency to gum. The oil holes in the bearings should be cleaned out by a wire, and the wearing surfaces of the slide bars should always be kept clean and the wicks in the oil cups with one end down the tube. All the oil cups should be filled with oil before starting, and they should be refilled every three or four hours at the furthest during the time the engine is working. Where glass lubricators are used care should be taken to see that they are properly adjusted and dropping freely. If any of the bearings have recently been tightened up and run warm they must be oiled oftener, and in case a bearing gets hot and the application of oil does not succeed in cooling it, ease the brasses slightly, but not to cause knocking in work. Should this not prove effectual, pour cold water over it, and in case it is found that the surface of the bearing or the shaft is injured, it must be smoothed over with a fine file where it is visible that an undue friction has existed.

Note.—Intelligible directions for filling and cleaning lubricators can only be given with the aid of diagrams. The operations, however, are very simple.

Starting the Engine.—Before starting carefully examine all the working parts of the engine, and if it is a quick speed governor engine, also the governor belt, to make sure it is in good condition and not liable to break before the work is done; also turn the engine round a few times by the fly wheel when practicable, having the cylinder drain taps open, the drain cock on jacket should also be opened to allow any water that may have collected to escape. Undue tightness in any of the bearings will be ascertained by moving the fly wheel backwards and forwards a few times, and if the straps on the eccentric sheaves are too tight the rods will vibrate ; and the straps will be eased a little by unscrewing the bolts a half turn or less, taking care that the lock nut on the opposite side is again tightened up. These points should have careful attention before admitting steam into the cylinder, and on no account should steam be admitted by at once opening the valve full ; by doing so the steam pipe may leak at joints from **sudden expansion.**

The valve chests and cylinder should be gradually heated up to within a little of its working temperature by opening the valve slightly, and on the valve being opened a little more the engine will move slowly. After it has worked for a short time and steam pipes, valve chest, and cylinder got properly heated up, the valve may be opened full, when the engine will run at full speed. When the cylinder taps have discharged all the water and emit nothing but dry steam, they may be closed, as also the cock on jacket If the engine is a considerable distance from the boiler, the steam pipe should be properly drained before reaching the stop valve of the engine.

The following remarks are intended for Marshall's Engines, but they also apply to many others.

Feeding Boiler and Management of Feed Pump.—The force pump is constructed for continuous action ; that is, when the engine is running it is continually raising water from the feed tank ; and as its capacity is always in excess of the requirements of the boiler, only a portion of the water raised enters the boiler and the remainder is returned to the tank down the overflow pipe attached to the third or delivery valve box of the pump.

Immediately the engine has been started it is necessary to ascertain that the pump is working properly, and this is done by turning the handle of tap on the pump straight down, when water should flow copiously from the bottom end of overflow pipe, and in such event the tap on the pump must be partially closed. This tap regulates the course of the water raised by the pump ; when the mark seen on end of plug is in a horizontal position, all the water lifted is forced into the boiler, and when placed vertically or straight up and down, none of the water enters the boiler, but the whole is returned to the feed tank down the overflow pipe. In general work it will be found best to adjust this tap between the two abovementioned positions, so that the necessary water is forced in the boiler to maintain it at a uniform level, and the rest returned to the feed tank down the overflowing pipe. This will quickly be ascertained after a short experience in the management of the engine, and by its observance a regular pressure of steam may be more easily maintained than if the water is pumped into the boiler irregularly.

In case water fails to flow down the overflow pipe directly after starting the engine under the circumstances before stated, or the pump does not maintain the necessary quantity in the boiler during work, some obstruction exists that requires removing, and the pump may not work from one or more of the following causes.

1st. The holes in the bottom of globular rose at bottom of suction pipe closed up.

2nd. This globular rose only partially covered with water in the feed tank.

3rd. The internal bore of the suction pipe obstructed.

4th. Suction pipes leaking at air joints or somewhere in the length of the pipe.

5th. Pump valves stuck or become dirty.

6th. Pump leaks water around the plunger and requires re-packing, or the existing packing compressed by screwing down the gland.

7th. Internal bore of overflow stopped up.

8th. Passage between check valve and boiler obstructed.

Nos. 1, 2, 3, 4, 6, 7 may quickly be detected

and easily remedied. No. 5 will be most likely to occur when dirty gritty water is used and a very slight blow on the side of the valve boxes with a small piece of wood frequently sets them right ; at all events it should first be resorted to, and in case it fails, the overflow pipe should be taken off the pump and the handle of the regulating tap on same turned straight down ; if hot water then issues out where the overflow pipe has been attached, the third or delivery valve in immediate connection with the boiler is wrong, and if it cannot be put right with slight blows on the valve box with wood as above explained, it will be necessary to take out the fire and blow off steam to admit of the cover of valve box being removed, and the valve taken out and cleaned. This valve box cover must not be taken off whilst the engine is under steam, or all the water would be blown out of the boiler. The first valve above the suction pipe is the most liable to get fast ; the cover of this valve box may be at any time removed, the valve being taken out with the fingers and washed, the valve seating should also be washed and the pump filled with water ; the valve may then be put back again and the cover tightly screwed down, and the engine again started. Should the pump not then work properly, the second valve may also be taken out and cleaned in the same way as the first or suction valve *providing* hot water does not issue from under side of pump after the overflow pipe has been detached and the handle of the regulating tap placed straight down, as hereinbefore described ; but should hot water so issue out of the pump it is clear that the third or delivery valve is hung up, and if the cover of the second valve box were under such circumstances taken off, the water would be blown out of the boiler.

No. 8 Can only be made right when the engine
is not under steam, and is done by taking out the
small brass plug directly opposite the point at
which the water is delivered into the boiler, and
inserting a piece of strong wire, chisel pointed at
one end, thereby removing any obstruction that
may have become deposited in the passage way
for the water ; when this has been done put back
the brass plug and screw it tightly up.

Bearings.—If any part should heat, it should
be known to the attendant whether it is occasioned
by want of oil or the bearings being too tight.
Should it arise from want of oil, the oil hole must
be cleaned out and the cup refilled, if through
bearings being too tight they must be slackened a
little and oiled.

At the earliest opportunity take out the shaft or
whatever has become hot, and file the journal
smooth with a smooth file, also file the inside of
the brass with a half round smooth file ; the journal and brass must be filed so that the file marks
will be round the shaft, that is in the direction in
which it revolves.

General.—The boiler is sometimes said to
prime. This is said to occur when the water boils
over, as it were, and comes into the engine with
steam. This is not as a rule very dangerous to
the boiler, but it is to the engine, as the water
getting into the cylinder often causes a break down.
Priming is caused by using dirty water, by allowing
it to get too high in the boiler, or by irregular
firing. The cure is indicated by the causes. As
remarked before, the boiler should never be ex-
posed to sudden changes of temperature and
pressure, as these are apt to loosen the joints,
rivets and tubes. For this reason steam should be

got up slowly, and when the day's work is done, should not be blown off but allowed to cool and condense in the boiler.

Stuffing boxes are a fertile source of leakage ; when they are allowed to go any length of time without the packing being renewed, no amount of screwing up will make them tight. When they are observed to leak the gland should be tightened, care being taken not to tighten it so much that the friction of the packing on the rod will cause the latter to heat, and also that the nuts of the bolts be tightened evenly all round.

Should the leakage not cease when the gland is moderately tightened, the packing must be taken out and replaced by fresh packing. Care should be taken that the whole of the packing be removed from time to time. Any pieces that are not much worn can be put in again, but the new pieces should be put in first at the bottom of the stuffing box. When being put in all packing should be cut a little shorter than will go round the rod, so as to allow it to expand lengthways when screwed up. If cut with a bevelled end so that the ends overlap, so much the better ; the cut end should be at opposite sides alternately, so that steam cannot pass. Rope or cotton packing should be well soaked in tallow or oil.

Doors between Boiler and Engine room and the factory should always be kept closed to keep heat from entering the latter, and dust, &c., from entering the former.

Sec. 3. Oil Engines.—Petroleum Engines are now used in some Factories in Ceylon, so I give below instructions for working three kinds of them.

Tangye's Petroleum Engine, Pinkney's Patents. —Before starting see that the engine is free from grit and thoroughly oiled, also that the water is supplied to the Cylinder jacket and that the valves bear freely on their respective seats.

Ignition Apparatus. To start lamp probe the small hole in burner and clean the coils with wire brush, fix it beneath the Chimney over Ignition tube, then put several small loose pieces of asbestos in the cup beneath the coils of the lamp, pour some petroleum (Kerosine) in the cup and light it ; this will burn and heat the coils for vapourizing the oil for the lamp, then open the cock at the bottom of lamp, extinguish the flame in cup and the lamp will burn with a good steady flame. When the flame of the lamp gets low, probe the hole in burner with the tool supplied ; one end of this tool forms a case for the needles (size of needles No. 13.)

Each night after engine has finished working put the lamp out and open drain cock stamped "B" fixed at bottom side ; a small tin should be used to catch the oil.

Oiling. All working surfaces must be oiled as in an ordinary steam engine, with the exception of the cylinder, which only requires a few drops at the start. Tangyes Gas Engine oil should be used.

Starting and Stopping.—Having lit the lamps and allowed them to burn for ten minutes beneath the vapourizer and ignition tube respectively, put the relief lever which is at the end of the exhaust lever into the position where the small relief cam fixed by the side of the large exhaust cam will depress the exhaust lever, and ease the compression when turning the engine by hand. Turn on the petroleum supply cock near the tank to position marked " starting," and give the engine a few sharp turns, it will then commence to work. Now turn the cock full on and move the **Relief Lever** out of the way of the relief cam, and then turn out the lamp beneath the vapourizer only ; move the lamp under the ignition tube into position as indicated between the lines on the side of lamp bracket for full load, and closer up to the vapourizer for light loads. The small cover on chimney over ignition tube should be full open for full load, and closer up to the vapourizer for light loads (a little practice will soon determine the position of lamp and cover or covers for light loads). Always keep the lamp beneath the ignition tube burning when engine is working or required. Always stop the engine with the crank at the **Bottom Centre** on the **Compression Stroke.** The valves will then be closed.

Start the engine with the crank at the **Top Centre** on the **Firing Stroke.** The letters on the Side Shaft (viz., top when starting, will also be on the top side).

Do not place your foot on fly wheel when starting.

Filling Oil Tank.—First screw the plug stamped A (fitted in side box) tightly up, then remove filling plug at top of tank, fill tank, pouring the whole of the oil **through the Strainer,** supplied loose for cleaning purposes (this is important), replace plug tightly, then open plug stamped A and supply is ready for engine.

See instructions cast on top plug and lid of side box.

Water Tanks and Cooling.—Never work the engine without having water in the jacket of the cylinder ; the water in the cylinder jacket should not become much hotter than the water in the tank. The supply to the tank should be supplied by ball cock.

Cleaning.—To clean the cylinder or replace piston ring simply disconnect the crank end of connecting rod and draw the piston out, clean out any deposit that may be found at explosion end and then thoroughly oil the cylinder. Draw the piston say once in three months. In replacing the piston be sure to place the gaps in the ends of the rings round the steady pins at the bottom of the piston and so prevent breakage. When the piston requires a new ring, change the one nearest the explosion end first. Examine the exhaust valve say once in three months, and grind it on its seat with a little flour emery if necessary.

The Vapourizer should be cleaned twice each week ; to do this disconnect **oil tube,** draw **air pipe** out of **valve box,** take off cowl over top of **valve box,** uncouple valve box from vapourizer, take vapourizer off **back cover,** then clean all the burnt oil off the inside of vapourizer with the tools supplied. Clean out any small pieces of dirt (with tool also supplied) that may fall into **ignition tube** when clearing vapourizer, after which replace all the parts.

The Back Cover should be cleaned once each week ; to do this when then the vapourizer is off, there is also a special tool supplied with the engine for the purpose.

Side Shaft-gear.—Be sure that the marks A and A on the teeth of the gear wheels come opposite each other when putting together or the engine will not work properly.

The Hornsby Ackroyd Patent Safety Oil Engine is also used in some Factories. In order to start the engine a lamp is used to externally heat the vapourizer, which can be readily done in from five to seven minutes in the smaller sizes and proportionally longer in the larger, with the help of a small rotary fan attached to the engine. *The lamp is then extinguished,* as there is no further use for it inasmuch as the required heat in the vapourizer is kept up by the continuous explosions. The engine is kept running by the pump supplying a suitable quantity of oil, the supply being controlled by the action of the governor. Very little attention is then required beyond replenishing the self-acting lubricators periodically. Full instructions are sent with every machine for working it, but it is said to be so simple that any unskilled person can readily understand it.

The Campbell Oil Engine.—Some of these are also in use in Ceylon Factories. In heating the vapourizer of the Campbell Oil Engine no use is made of either 'a fan or a hand pump ; an oil lamp is lit by a match, and it heats the vapourizer in a few minutes, after which the engine is ready for starting. The heating of the vapourizer is maintained independent of the heat from the explosions in the cylinder, and on this account with light loads no attention has to be given to assist the heat of the vapourizer. The consumption of the oil is automatically regulated by the governor to suit the load on the engine maintaining an even speed. When not running with variable loads the governor cuts off the supply of oil when not required for the work.

Sec. 4. Fans.—The fans in general use in Ceylon are the Blackman's, Wings Disc, The Aland and Scott's Challenge. They are all worked in much the same way ; all that it is necessary to do is to turn them on and off as required with the belt striker, to keep them clean, and the bearings well oiled. Actual experience of the conditions of his own Factory is the best guide to the Tea-maker as to when and for how long the Fans should be turned on.

They are generally driven at about 600 revolution a minute.

Sec. 5. Rollers.—The rollers in common use in Ceylon Factories are the Economic, the Rapid, the Triple Action and the Sirocco, and a short description of the methods of working them is now given.

Economic Roller (Jackson's).—The charge for this roller is from 100 to 120 lbs. of withered leaf. It is generally sent down a shoot from upper floor of Factory into a hopper on the box of the Roller; the leaf should be introduced gradually, the machine being started in motion when it is about a third full and the rest added as the leaf gets rolled and the machine can absorb it. Any leaf thrown out on the table by the box when revolving should be swept back with a hand brush.

Accidents often occur when the Roller coolies try to push back the leaf with their hands.

Pressure is applied by means of a hand screw on a weighted cap fitted on a davit above the box. All bearings should be well oiled before work commences, and when the day's rolling is finished the machine should be thoroughly cleaned and washed, the weighted cap on davit can be swung out to facilitate this. Amount of pressure and time leaf should be rolled is discussed under heading "Rolling" given before. I find the machine does the best work when it is charged with not more than 100 lbs. of withered leaf and box is driven at 60 revolutions per minute.

Large or 28 in. Economic Roller (Jackson's).—The general principal of this Roller is the same as the small Economic except that it is larger and takes up to 200 lbs. of withered leaf. The remarks above also all apply to this Roller.

Rapid Rollers (Jackson's).—The large size has a 32 inch box, the small size a 24 inch box. The former is in two types, *i.e.* with a square box, and a circular box. They are all worked in the same way, and the only thing in which they differ is the amount of withered leaf they can roll at a time. The capacity of the 24 inch and 32 inch Rollers is stated by the makers to be 200 and 300 lbs. of withered leaf respectively, but I find in practice that they do better work if they are only filled with 180 and 250 lbs. respectively.

The method of working is the same as with the Economic Roller except that pressure is applied by means of chains running over pulleys at the side of the machine, which are pulled up and down by the Roller cooly.

Speed should be 60 revolutions of box per minute.

Triple Action Roller (Brown's).—This machine is made in three sizes : large, medium and small. The makers claim that they can roll 475 lbs., 350 lbs., and 175 lbs. of withered leaf respectively at a time. I do not know the large and small types, but in the medium sized which is in general use, I find that about 250 lbs. of withered leaf is the best fill, though it is possible to get 300 lbs. in the Roller.

The leaf is generally put in the Roller from the upper floor of the Factory down a shoot ; the machine should be started when it is about a third full, and the rest of the fill added as the machine revolves. Pressure is applied in some of these Rollers by means of a weighted lever which is easily manipulated, in others by a hand wheel above the box. The leather belt driving the upper rolling surface should be twisted. All bearings and slide surfaces attached to the frame of the machine should be kept well oiled, and the roller washed down and thoroughly cleaned when rolling for the day is finished.

The machine should be driven at speed—(box) of 48 to 50 revolutions per minute.

Sirocco Roller (Davidson's).—The method of working this machine is very simple. Withered leaf is sent down a shoot from upper floor in the usual way : the Roller can be started in motion as soon as the leaf begins to descend. There is no means of applying pressure. The makers say that it takes a charge of 350 lbs. of withered leaf at a fill, but I find that it does better work when only filled with from 250 lbs. to 300 lbs. of withered leaf. The pulleys on the machine should be driven at a speed of 150 revolutions per minute.

Bearings should be kept well oiled and the machine should be washed down and thoroughly cleaned when work for the day is finished.

Sec. 6. Roll Breakers.—Those in use in Ceylon are so very simple that no description of their methods of working is required here. As with all other machines, the bearings must be kept well oiled, and the machine, especially the mesh of the sieve, clean.

Sec. 7. Davidson's Tea Oxidizer.—Some of these are now in use in Ceylon Factories. A fill from a large Roller, *i. e.* from 250 to 300 lbs. of withered rolled leaf forms a suitable charge for the machine. It is used for chilling and oxidizing

the rolled leaf after it comes from the Roller, and
also for evaporating a portion of the moisture
contained in same, The leaf in some Factories
is put in the machine almost as soon as it leaves
the Roller ; in other Factories, after it has been
fermented, when it is said to instantly fix the
colour of the ·leaf, and the chilling it receives to.
temporarily check further fermentation prior to
the leaf being passed into the Drier.

As each tray full of leaf is placed on the top of
one of the compartments, the valve handle is drawn
forward, which opens the valve port (leading to the
exhaust fan) and allows the Fan to draw the air
away from below the tray. The time the leaf
should remain on the trays of the machine varies
with the condition of the leaf, &c., and can only
be decided on after some experience in the working of it in each particular Factory.

McDonnell's Cooler and Oxidizer is also in
use in some Factories, and consists of a blower or
other air supplier connected by means of a pipe
with the receptacle or box of a Roller. I do not
know that any special instructions are required for
working it.

Sec. 8. Driers.—The following are the Driers
in general use in Ceylon Tea Factories :—Sirocco,
Desiccator, Venetian, Victoria and Paragon.

Siroccos: (Davidson's) Up-draft.—There are
two types of the Up-draft Sirocco, *i. e.* the End
Slide, and the Side Drawer. In the first the trays
are filled with leaf and pushed through the machine
from end to end. In the second the machine is
worked from the front, each vertical row of trays
being worked independently.

When the fire is being lighted preparatory to
working the apparatus, a small bundle of straw

4

or shavings should be lighted and passed into the smoke box at base of chimney, and allowed to burn there freely for two or three minutes before lighting the fire itself. This is for the purpose of warming the chimney, and starting the draught.

The air valves along the trays should at the start be closed completely to prevent the escape of air up through the empty drying chamber until the fire is burning freely, and the stove at a good heat, when they may be gradually opened. **In End slide Siroccos,** the apparatus should be filled with empty trays, and when the temperature as indicated by the thermometer on machine is at the right height, a tray of wet roll should be inserted at one end of top, and the tray pushed out at the other end and inserted in lower slide, and so on right through the apparatus ; and proceed by putting fresh trays filled into top slide and shifting the trays one move down, until the bottom space is reached, when the finished tray is withdrawn ; this rotation movement and replacement of fresh trays on top is repeated until the leaf is all fired. As the leaf arrives at each of the slide ends from the other it shuld be turned over on the trays.

The rolled leaf should be spread in moderately thin layers on the trays ; thinly charged trays and prompt quick pushing along of them will do considerably more work than can be done by spreading the trays thickly.

With Side Drawn Siroccos the method is much the same except that a tray of wet roll is put into the top row of each of the vertical rows of slides, and empty trays into all lower slides, then after a short interval, proceed in the ordinary way by extracting the top tray, turning over the leaf on it, and inserting in second row, and so on one move down until the bottom is reached ; after being in bottom row a short time, the tea on being withdrawn should be found fully fired.

Regulation of Temperature.—The temperature is regulated mainly by the supply of fresh air which is drawn from below the fire. The sliding air valves which are fitted in the doors on the openings into the ashpit from the back and front plates, control this supply of air, and the cooly, by moving both or either of the valves to one side or other, can regulate the temperature of the apparatus to any required degree. Working temperatures have been remarked on in paragraph " Firing" given before.

Stoking.—These remarks apply to the stoking of all the various Firing Machines described in this book.

When wood fuel is used it should be split into small pieces and added to the fire frequently and in small quantities, the fire being kept about 6 to 8 inches deep. The draught should always be of sufficient strength to enable full heat to be maintained without having to poke or rake the fire frequently, as a fire that has to be often stirred up gives bad results for the amount of fuel consumed. If the draught be poor either the chimney or flues requires cleaning. These should, however, be thoroughly cleaned out every morning before commencing work. When coal fuel is used the fire should on no account be kept more than 3 or 4 inches deep on the bars, and fresh coal should be thrown on the fire at frequent intervals in small quantities at a time. Thick deep fires should not be used, as a loss of fuel is caused by the imperfect combustion of the gases in the coal which pass away up the chimney without ignition ; and the fire bars are apt to get red hot throughout from deficient air current up through them and sag down, and all the iron of the furnace will wear out much quicker than if thin fires are used. When using coal fuel the fire requires to be cleaned about every four hours—that is, the bars well scraped with the scraper and all the clinkers removed.

Siroccos Down-Draft (Davidson).—There are two types of these, though each is made in various sizes. The first fires tea in the ordinary way, the second in the so-called "inverted" way.

Ordinary Down-Draft Sirocco.—Method of working. Light fire and stoke in some way as described in Stoking in, Up-Draft Sirocco. The damp rolled leaf should be evenly spread on the trays, care being taken to break up thoroughly any lumps in it, and to spread it level on the trays. The leaf should on no account be pressed or patted down. One man is required to put in and take out the trays from the drying chamber, and another to prepare and spread the leaf on them.

When the proper temperature is attained, as indicated by thermometer on machine, take a filled tray, and resting one edge of it on the ledge of the lower tray Port, raise the handle or lever of the tray-lift to the top of the guard. This simul-taneously lowers the tray lift, opens the lower tray port door, and shuts the fan valve. The tray is then pushed into the drying chamber, through the open tray port, and the lever pressed down again, whereby the tray is raised on the tray lift past the lower tray pawls or catches which then fall in below it and prevent it going down again ; while at the same time the tray port door closes and the fan valve re-opens, which allows the full suction power of the fan to operate on the whole bottom surface of the tray ; and it must be noted that the tray lift handle must always be kept opposite the plate marked "Fan valve open" except when a fresh tray is being inserted. The strength of the fan suction is such that it draws the hot air from the top of the drying chamber down through the leaf with great velocity, but notwithstanding the great force of the blast no leaf gets disturbed on the tray. After a half minute's interval the second tray is inserted in the same way as the first, and on its being similarly elevated, it lifts the first tray on top of it, and takes its place on the lower tray pawls. With intervals of half a minute between the insertion of each tray, this operation is repeated until the drying chamber contains a column of six trays, the top one of which will now be exactly opposite the intermediate tray port, through which it is withdrawn, and the leaf on it shaken up and re-spread, and the tray pushed back to its place. A fresh tray is then introduced at the lower tray port in the same manner as above described, and on the lever being pressed down it will again lift the column of trays, and in doing so the top one will be raised past the upper set of tray pawls on which it will be supported when the lever returns to working position, thus leaving the tray below the "upper pawls" free to be withdrawn and the tea turned. On this tray being reinserted the same method is followed till there are (in the large machine) ten trays in the drying box when the top one will have reached the upper tray port door, where it is again drawn out and examined, and if the tea appear thoroughly dried, it may then be removed from the machine. If, however, it is uneven and any spots appear damp, it should be shaken up and put back for half a minute longer, when it is withdrawn and a fresh tray of damp leaf inserted at the bottom, so as to always keep the drying chamber filled. The manufacturers advise that the teas in these machines should be fired at a temperature of 260° but from actual experience I find this is too high for Ceylon, as the tea is liable to get over fired or even burnt. Suitable temperature is, I think, 220° to 235° and certainly not above 240°.

Method of Working Down Draft Siroccos fitted for Inverted System of Firing.—By which the wet roll is subjected to a high temperature at the beginning to arrest fermentation, is about half fired and then finished off at a lower temperature.

When the machine has attained the required temperature, take a tray on which the damp roll to be dried is spread, and insert this tray firstly into the **top** tray port where it should be retained for about two minutes. This tray should be then drawn out and the leaf turned over, shaken up and re-spread on the tray, after which raise the handle or lever of the tray lift to the top of the guard. This simultaneously lowers the tray lift, opens the lower tray port door, and shuts the fan valve. The tray containing the now partially dried leaf should then be inserted into the lower tray port, and the lever pressed down again, whereby the tray is raised on the tray lift past the tray pawls or catches which then fall in below it and prevent it going down again, Another tray spread with damp roll is then introduced into top tray port and allowed to remain for about two minutes. The tray should then be drawn out the lever raised to the top of the guard (and kept there until a fresh tray of wet leaf is inserted,) the leaf is then to be shaken up and re-spread as before, and another tray of damp roll put into the top tray port and lever lowered. This operation is repeated until the lower chamber is full of trays, the top one (of the lower chamber) is then examined, and if the tea is found fully dried, the tray is removed from the machine. A convenient way of working is to have a tray of wet roll ready spread in advance, and as the one containing the partially dried tea is removed from top port, to at once insert the other into the same port. The lever is then lowered and the first tray turned over and respread. **It is most · important to note that when there is no wet roll in the top port, the lever must be kept up, and the fan valve closed or shut off.** The insertion of the tray of damp roll into the top tray port, and allowing it to remain during the interval between each change of tray below has the effect of instantaneously arresting further fermentation of the leaf owing to the sharp heat of the air coming direct from the stove, and also fires it from about a third to a half. The leaf from the tray to be withdrawn from the middle tray port will not feel excessively hot to the hand when withdrawn from the machine, owing to the air in which it was finished having previously passed through the tray of damp leaf in the top compartment, which reduces the initial temperature of the hot air by a great many degrees. From actual experiment I find that it is about 70 to 75 degrees, as with air at 235° in top compartment it is generally 170° in tray at top of column in middle compartment. The manufacturers recommend that the fan should be driven at 900 revolutions per minute. I prefer 800 for actual work. All the recent Down-Draft Siroccos have the exhaust nozzle of the fan fitted with a vertical slide valve or damper which regulates the temperature of the air passing into the drying chamber by controlling the volume drawn through the stove. If a low temperature be wanted the valve is lifted full open, and if a high temperature be required it is lowered step by step (hole by hole) till the requisite increase of temperature be attained.

Auto Siroccos (Davidson's).—I do not know that any of these are yet in Ceylon, and have not been able to obtain the methods of working them ; should, however, a second edition of this book be required and the machines by then have come into use in our Factories, I hope to supply the omission.

Note.—If final firing before packing has to be done in a Sirocco fitted to fire tea in the "Inverted" method the *lower column* of trays only should be utilised, no use whatever being made of the top port which should be kept closed. The trays filled with tea to be final fired should be put into the lowest port and drawn out with the tea finished at the middle one.

Desiccators (**Brown's**). —These driers are made in three sizes, Nos. I, II, and III., and they are all worked in much the same way.

No. I machine consists of a furnace and drying chamber, the latter having one tier holding four trays and a space at one end for one finishing tray. The trays charged with the leaf to be dried are entered into the drying chamber at intervals of a few minutes and pushed through to the other end the time between the insertion of each tray depending on the temperature employed and the speed at which the fan is driven. The actual time required is easily known after working the machine for a short time. Trays should be thinly and evenly spread. If the tea on the trays on being withdrawn at the finishing end of the drying chamber is found to require further firing, it is put into the extra firing space to finish off.

In No. II and III machines the drying chamber is divided into two complete or independent parts, each having its tier of four trays, and one extra space for finishing. Each chamber is supplied with a separate current of heated air direct from the furnace. This current of air can be regulated by a valve at the side of the apparatus worked by a handle on a rod from the filling end of the machine. A little experience will soon show the Teamaker how to thus regulate the temperature and to keep it exactly as wanted.

The trays filled with the leaf to be dried are generally put into the drying chamber from the end furthest from the fan on the top tier, and are pushed through one after the other successively, when they reach the other end the cooly there turns over, shakes up and re-spreads the half fired tea on them, and inserts them on the lower tier ; they should arrive at the starting end but in the lower tier fully fired, but if not quite so they can be put into the extra firing space, which is at this end to finish off. Fan should run between 400 and 500 revolutions per minute. Rules on firing and stocking have been already given, the latter below paragaraph on Up-Draft Siroccos.

It is quite possible, if desired, with Nos. II and III Desiccators to keep the temperature of top tier of trays at 230° to 235°, and the lower at 180° to 185°, thus firing the teas on the "half firing" system. Fan must be kept well lubricated and interior of machine cleaned out daily.

Venetian Drier (Jackson's).—In this apparatus there are five drying surfaces in the chamber, over which is mounted a powerful exhaust fan, which induces air through the furnace tubes ; the air is heated on its passage and is drawn *upwards* through the tea leaf which is spread on perforated plates or trays. The top series of trays are mounted in an iron frame which serves as a feeder, this is pulled out to receive a charge of leaf and pushed in to discharge it on those underneath.

Method of working.—A charge of damp roll is placed on top tray which is pushed into chamber, and a disc revolved quickly to discharge the leaf on to top tier of trays in drying chamber. This operation is repeated until all the five tiers of trays are fully charged; care being exercised to move cranks in their turn, so that two charges of leaf shall not accumulate on one tier of trays. When fairly started the crank handles are then turned consecutively beginning at the bottom, when each stratum of tea will be moved down from one tray to another, the lowest being delivered from the machine chamber by a shoot direct into any convenient receiver.

Fan should be driven at 550 revolutions per minute, and its bearings kept well oiled.

Stoking.—See remarks under this heading (paragraph Up-Draft Siroccos).

Temperature.—See remarks under paragraph " Firing."

Victoria Drier (Jackson's.)—This is an automatic or self-working machine, and the drying is effected by the leaf being mechanically and continuously fed on to the endless metallic perforated travelling webs placed above each other, which carry the leaf through the drying chamber and automatically deliver it dry outside.

The firing cooly places the wet roll to be fired into the feeder at the top of the machine from which it is automatically scattered on to the uppermost web. It is however possible, if desired, to scatter the leaf by hand. The machine has five slowly moving webs inside, over which the leaf is effectually turned four times during the process of drying. The leaf to be dried can be passed through the machine at five different rates of speed at the will of the Tea-maker ranging from 9 to 25 minutes.

Air is drawn **upwards** through the leaf by a powerful exhaust fan. Speed of fan should be about 320 revolutions per minute. For working temperature see remarks under paragraph "Firing" given before. For Stoking see remarks in that paragraph under Up-Draft Siroccos.

Empress Drier (Jackson's.)—The principle of drying by this machine is much on the same lines as that of the Victoria Drier, but it has no mechanical feed regulator, the leaf being simply scattered by hand on to the upper web which projects beyond the end of the machine to receive it. The leaf is turned over five times whilst in the drying chamber.

Speed can be varied as in the Victoria Drier.

Speed of fan should be 375 revolutions per minute.

Working temperature and stoking, see remarks under Victoria Drier.

Paragon Drier. (Jackson's).—In working this machine the wet roll is fed or poured into the web by a cooly standing on a platform fixed to the front of it which also covers the discharge. A vibrating plate which has a vertical adjustment regulates the thickness of the feed, and if at any time it should be necessary the leaf can be scattered by hand on the slowly moving web. The leaf can be carried through the drying chamber in ten minutes, or this time may be increased to 25 minutes by changing the strap on to the different steps of the cone pulleys.

For final firing before packing, when it is desired to do this slowly and at a low temperature the inlets to the fan can be shut down and the webs run at their slowest speed. A large door fitted with asbestos is fitted on each side of the machine, and one on the top, so that the whole of the internal working parts can be examined at any time. The inner end of the tubes in the stove is also accessible through the side doors in the drying chamber. The air is drawn through the moving tea by a powerful exhaust fan which should run at 450 revolutions per minute.

For Temperature see remarks in paragraph " Firing."

For Stoking see remarks under that heading in paragraph Up-Draft Siroccos.

Chota Paragon (Jackson's).—This is a small sized Paragon. The drying chamber is placed on the top of the stove. The travelling webs have five different speeds. It is worked in the same way as the larger machine, and speed of fan is the same.

The tubes are all get-at-able on their outer and inner surfaces, and can be brushed and cleaned at any time. It is essential that all Driers should be kept thoroughly clean inside and out, and all bearings well oiled and the lubricating bottles filled.

Sec. 8 Sorters.—The ones in general use in Ceylon are Walker's, Brown's, Jackson's, Eureka, and Davidson's. The working of them all is very simple, and detailed methods are not required here, as every Superintendent has his own rules for sifting, which Tea-makers have to carry out. The three first mentioned Sorters have generally inter-changeable trays of various sized mesh, so that the tea can be graded with as little handling as possi-ble into any desired sizes or varieties. The Eureka is fitted with cutters or " reducing mills " attached to and driven by machine. The top one cuts the tea for sifting, and also serves the purpose of a mechanical feeder. The other equalises the large tea that will not pass through the top mesh or sieve. The "reducing mills" can be readily thrown out of gear and the sifter worked without them if desired.

At the top of the Davidson's is a fluted feed roller with an adjustable resistance plate at the back of it. This either cuts or equalises the tea as desired, as it can be set close to or far apart from the roller by means of an adjustment screw on the hand lever attached to it.

Sorters should always be thoroughly cleaned after work.

Sec. 9. Cutters.—These are so very simple that no instructions for their use are required here. In some Factories only the coarse leaf and pekoe souchong are cut, in others the pekoe is also passed through the machine. Less dust is made when the Cutter is driven slowly. The dust should be removed by a hand sieve (No. 30 or 32) after the tea has passed through the cutter, or if flat pieces are also required to be removed, by a No. 24 sieve.

Sec. 10. Packers.—The machine in most general use in Ceylon Factories is the Davidson-MaGuire Tea Packer. I believe that Jacksons are bringing out a new tea packing machine, but it has not yet been introduced into Ceylon. Should it come into use here, and another edition of this book brought out, the method of working it will be described then.

Davidson-MaGuire Packer.--Method of working. The empty chest to be packed is first of all secured firmly on the packing table of the machine by the self centering clamps, which are worked from the side by a hand wheel, and secured from slacking by a slight movement of a lever clamp also at the side, which tightens up the bearing on the screw spindle of the clamps.

At the back of the table there is a stopper in the form of an adjustable screw with a button end, which can be set as required for instantly regulating the distance to which chests of one uniform size are to be pushed back on the table, so that the centre of each *lengthwise* may be fair on top of the clamping screw, before the clamps are tightened against the ends of the chests.

After the chest has been firmly clamped to the vibrating table, the top sides of the lead should be secured to the edges of the box by clips, one on each side (it is well to have a piece of brown paper inside these to prevent them tearing or piercing the lead) ; the hand gear on the striking gear of the countershaft is drawn to start the machine. The tea has then merely to be thrown into the chest in large shovels of about 6 or 8 lbs. at a time with only four or five seconds' interval between each shovelful. Each shovelful should be emptied into a different part of the chest from the one before, into the corners first, then along the sides and in the middle. If done in this way it is unnecessary to touch the tea by hand at all. Coolies can be very quickly taught to fill the chests in the way described. Small tea such as Broken Orange Pekoe and Broken Pekoe should be thrown in expertly and quickly, as it packs so tight that the chests will hold more than 100 lbs. nett, if thrown in slowly, which is undesirable. Whole leaf teas, Orange Pekoe, Pekoe, and Pekoe Souchong should be thrown into the chest at a slower rate and with a longer interval between each shovelful to allow them to pack closer, as size for size they are proportionately lighter than broken teas.

PART IV.

USEFUL NOTES FOR TEA-MAKERS.

Venesta Chests.—These are used in some Factories, and I should like to have given full instructions for putting them together, but find it is impossible to do this intelligibly without the aid of diagrams, which are beyond the scope of this book. Full instructions with diagrams are, however, supplied by the manufacturers of the chests when purchased from them or their Agents.

Tea Lead :—

Cases of 2 cwt. 84"×22" of 5 oz. lead contain 56 sheets.

,,	,,	2	,,	84"×22"	,,	4½	,,	,,	,,	64	,,
,,	,,	2	,,	84"×22"	,,	4	,,	,,	,,	70	,,
,,	,,	2	,,	77"×22"	,,	4½	,,	,,	,,	60	,,
,,	,,	2	,,	72"×19"	,,	4	,,	,,	,,	94	,,

It is therefore easy for the Tea-maker from above table to check his issues of lead to the packing cooly by working out the length required for each chest used according to inside measurement and allowing 2 to 3 inches for overlap.

A new tea lead has lately been introduced in Ceylon called the "Silver Brand". The size is 84"×22", and it only weighs 2 oz. to the square foot. It is said to be tougher than ordinary tea lead, though thinner. I mention it to tell Tea-makers that as it is self-soldering, they do not require to use solder or soldering fluid to join it, a hot soldering iron and some rosin being all that is necessary.

Nails, Wire.—1 inch long, a pound should contain 656.

1¼	,,	,,	,,	,,	583.
1½	,,	,,	,,	,,	320.
1¾	,,	,,	,,	,,	295.
2	,,	,,	,,	,,	210.

Above table is useful in checking quantities used for a break of tea, the number in a chest being first counted.

Hoop iron ½ inch 24 B. W. G. 1,300 feet to the hundredweight.

Solder.—One stick of purchased solder is sufficient for twelve full chests.

Solder can be made with scrap lead and block tin mixed, proportions 2 lbs. scrap lead to 1 lb. block tin, the lead should be melted in an iron spoon or pot, and the tin added when the lead is liquid ; when the two have dissolved and mingled the mass should be stirred up (the dross skimmed off) and poured into grooves cut into a block of wood, the grooves being of the same size as an ordinary stick of solder. A pinch of sulphur added to the lead before it is hot will bring most of the dirt, &c., in it to the surface as soon as it melts.

Soldering Fluid.—After soldering down the top of a chest, all waste fluid should be carefully wiped off the surface.

Boxes.—A cooly can line with lead ready for packing 75 full chests or 100 half-chests in a day. This is an easy task, and an expert cooly can do more, as I know from actual experiment, provided he has a good stove (the Firefly is the best I know for the work) alongside him and three soldering irons. The stove I have mentioned, after it is once fairly alight, requires no attention, except to occasionally replenish it with *good charcoal*. It makes no smoke and does not require to be blown with bellows if the charcoal has been properly burnt.

Belting.—Leather belting should be occasionally rubbed over with fish oil. Cotton belting with linseed oil.

Belting paste is better to use than rosin if the belts do not cling tight enough to the pulleys.

Cotton belting expands very much in hot dry weather, but it should not be shortened as long as the belts will work, as it contracts again at once when the weather changes. It shrinks up if the air gets very moist.

When putting on a new or repairing a broken belt, the holes for the fasteners should be cut out with a punch of the exact size required, and not cut out with a chisel or knife.

A set of punches of sizes should be kept in all Factories.

Thermometers.—Should the tube of a thermometer break and a spare one be purchased, the latter should be tested after insertion in the frame, by putting it into boiling water. If it registers 212 degrees it is correct, but if not, the necessary corrections can be made on the scale in red or black ink.

Lamps.—When the burners or wire mesh get dirty, they should be boiled with common washing soda for a time.

Wicks.—Before inserting new wicks into lamps, they should be soaked in English vinegar, and then dried in the sun. This will help them to burn brightly and without smoke.

Lamp Chimneys.—Before taking a new Chimney into use, it should be boiled in salt and water. Chimney should be put into a receptacle containing just enough cold water to cover it, a half tea spoonful of salt added, placed on a fire, and boiled slowly for about an hour ; this will render the glass tough, and it will last longer than if not heated so.

Brass Work is best cleaned with Brooke's Soap or other efficient patent substitute, but should brass work have been allowed to become very dirty, the worst of the muck can be taken off by rubbing it with a mixture of the juice of a green lime and bath brick dust. The soap to be applied afterwards.

Bright Steel Work should be kept clean with a burnisher, emery paper, or soap ; if the latter is used it must be well rubbed with a dry cloth afterwards, and then apply the burnisher.

Waste Oil from the oil catchers, &c., can be used again after straining it through a piece of fine muslin.

Glass Windows can be cleaned and polished by rubbing them with a lump of newspaper dipped in water; the commoner the paper used the better, as it is the printing ink that effects the cleaning. The glass should afterwards be well rubbed with a dry cloth.

Spanners and Tools belonging to each machine should be kept separate, each set on a bracket or shield hung on the wall close to the particular machine it belongs to.

Lubricant for Wire Drive.—This is made from a mixture of $\frac{1}{2}$ lb. Tallow to three bottles ($\frac{1}{2}$ a gallon) of coal tar boiled together. It should be rubbed over the wire rope every other or alternate day the drive is at work.

Smoke Chimneys.—Small holes in these can be plugged up with a mixture of Portland cement and coal tar.

It is useful for Tea-makers to know the rules made by Government under Sec. 4 of Ordinance No. 2 of 1896, so they are given here.

In these rules, unless the context otherwise requires,—

"Child" means a person under the age of 14 years. "Young" person means a person of the age of 14 years and under the age of 18 years and upwards.

"Machinery" includes any driving strap or band.

"Mill gearing" comprehends every shaft, whether upright, oblique, or horizontal, and every wheel drum or pulley by which the motion of the first moving power is communicated to any machine.

(1) Every hoist or tackle, near to which any person is liable to pass or to be employed, and every wheel, if within a distance of 3 ft. 6 in. from the floor, directly connected with steam, water, electrical, or other mechanical power whether in the engine-house or not, and every part of a steam, gas, or oil engine, electrical motor, dynamo, or water wheel, shall be securely fenced ; and

(2) Every wheel race not otherwise secured shall be securely fenced close to the edge of the mill race ; and

(3) Every part of the mill gearing belting ropes or chains driven in any way whatever within a distance of 3 ft. 6 in. from the floor conveying motive power from the driving to the driven machine, shall either be securely fenced, or be in such position, or of such construction as to be equally safe to every person employed in the factory, as it would be if it was securely fenced ; and

(4) All fencing shall be of wood 3 ft. 6 in. high, post 3 in. by 3 in., top rail 3 in. by 2 in., intermediate rails two in number, 2 in. by $1\frac{1}{2}$ in.

If iron is used for rails it must be not less than 1 in. in diameter, and shall be maintained in an efficient state while the parts required to be fenced are in motion.

(5) All wires conveying the electrical current from the dynamo to the motors or lamps shall have highly insulated covering and be safe to every person employed in the factory ; and

(6) Every main switchboard shall be under lock and key and bear clear instuctions for its use by the inexperienced.

(7) A child shall not be allowed to clean any part of the machinery in the factory while the same is in motion by the aid of steam, gas, oil, electrical current, water or other mechanical power.

(8) A young person or woman shall not be allowed to clean such part of the machinery in a factory as mill gearing belts, ropes or chains conveying motive power from the driving to driven machine while the same is in motion.

(9) A child, young person, or woman shall not be allowed to work between fixed and traversing parts of any self-acting machine while the machine is in motion by the action of steam, water, electrical current, gas, oil or other mechanical power.

(10) No person shall have control over a boiler in which steam is generated for the purpose of driving machinery, unless he can produce a certificate of competency from the Inspecting Engineer appointed by Government, or from an Engineer specially licensed for the purpose by the Government.

Final words to Tea-makers.—They should always bear in mind that it is impossible for them to keep their machinery and factories too clean, that all work should be done systematically, that there should be a place for everything in the factory, and that everything should be in its place.

THE END.

www.ingramcontent.com/pod-product-compliance
Lightning Source LLC
Chambersburg PA
CBHW020237090426
42735CB00010B/1730